Why Living Today is Better than Yesterday

Illuminating Steven Pinker's Theory of Progress

David Christopher Lane

Andrea Diem-Lane

Mt. San Antonio College
Walnut, California

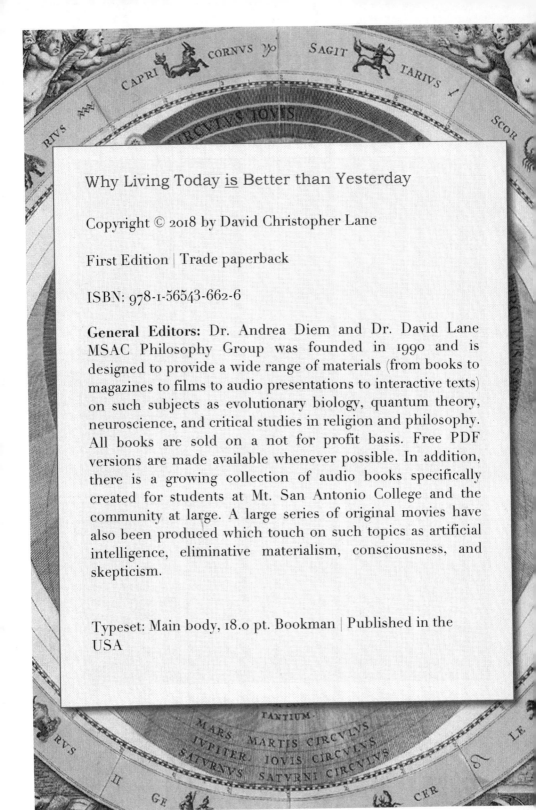

Why Living Today is Better than Yesterday

First Edition | Trade paperback

ISBN: 978-1-56543-662-6

General Editors: Dr. Andrea Diem and Dr. David Lane MSAC Philosophy Group was founded in 1990 and is designed to provide a wide range of materials (from books to magazines to films to audio presentations to interactive texts) on such subjects as evolutionary biology, quantum theory, neuroscience, and critical studies in religion and philosophy. All books are sold on a not for profit basis. Free PDF versions are made available whenever possible. In addition, there is a growing collection of audio books specifically created for students at Mt. San Antonio College and the community at large. A large series of original movies have also been produced which touch on such topics as artificial intelligence, eliminative materialism, consciousness, and skepticism.

Typeset: Main body, 18.0 pt. Bookman | Published in the USA

Dedication

To Charles Darwin who showed us how truly lucky we are to be alive.

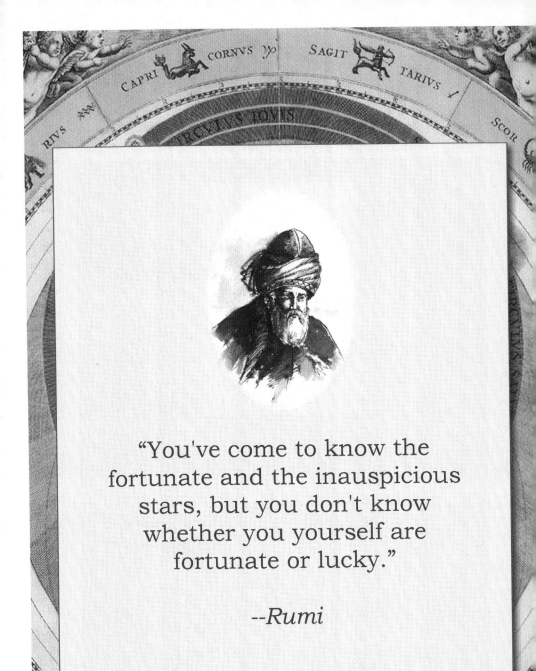

"You've come to know the fortunate and the inauspicious stars, but you don't know whether you yourself are fortunate or lucky."

--Rumi

Steven Pinker, the distinguished and widely known Professor of Psychology at Harvard University, caused an uproar in academic circles back in 2011 when he argued in his large tome (832 pages), *The Better Angels of Our Nature*, that the world is significantly less violent than at anytime in our history.

While Pinker's thesis has received accolades from such luminaries as Peter Singer, Michael Shermer, Richard Dawkins, and James Q. Wilson, it has also garnered its fair share of criticism, ranging from the anthropologist R. Brian Ferguson to the psychologist, Robert Epstein, who has each questioned parts of his data and methodology.

Pinker has now come out with a sequel entitled, *Enlightenment Now: The Case for Reason, Science, Humanism, and Progress* (Viking Press), which has garnered much attention and some off-the-chart reviews, with the cofounder of Microsoft, Bill Gates, calling it his "favorite book of all-time."

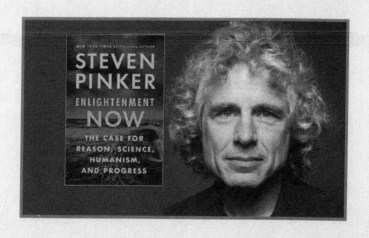

While I think it is important to keep our skeptical antenna finely tuned before channeling and accepting everything Pinker argues in his new book, I do believe there is overwhelming merit in *Enlightenment Now's* affirmative message.

"Human nature is complex. Even if we do have inclinations toward violence, we also have inclination to empathy, to cooperation, to self-control."

-- *Steven Pinker, Harvard University*

Indeed, Pinker's book may well usher in a new wave of self-help books that are based on comparative data markers ranging from medicine to sociology.

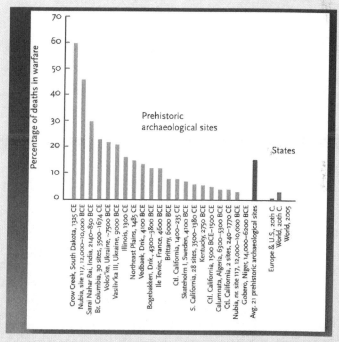

While Pinker may not necessarily agree with Gottfried Leibniz's theodicy that this earth is *le meilleur des mondes possibles* ("the best of all possible worlds"), he clearly believes that living right now is undoubtedly the best humans have ever had it on terra firma.

To substantiate this bold claim, Pinker presents 15 distinct metrics which underline his thesis that things have gotten much better for the general lot of humankind. Bill Gates cites five of his favorites in his widely cited blog, *gatesnotes*.

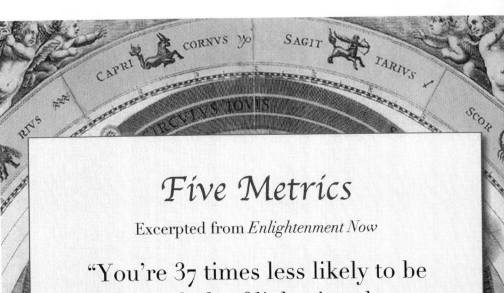

Five Metrics

Excerpted from *Enlightenment Now*

"You're 37 times less likely to be killed by a bolt of lightning than you were at the turn of the century—and that's not because there are fewer thunderstorms today. It's because we have better weather prediction capabilities, improved safety education, and more people living in cities."

CAPRI CORNVS
SAGIT TARIVS
CIRCVLVS IOVIS
SCOR
TANTIUM.
MARS. MARTIS CIRCVLVS
IVPITER. JOVIS CIRCVLVS
SATVRNVS SATVRNI CIRCVLVS

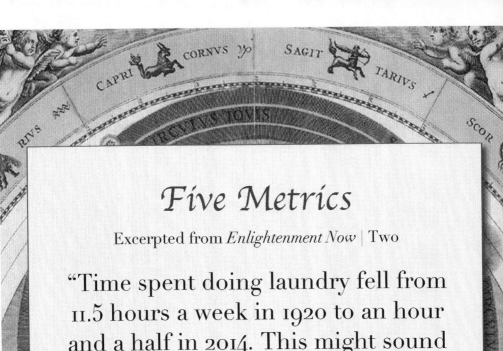

Five Metrics

Excerpted from *Enlightenment Now* | Two

"Time spent doing laundry fell from 11.5 hours a week in 1920 to an hour and a half in 2014. This might sound trivial in the grand scheme of progress. But the rise of the washing machine has improved quality of life by freeing up time for people— mostly women—to enjoy other pursuits. That time represents nearly half a day every week that could be used for everything from binge-watching Ozark or reading a book to starting a new business."

CIRCVLVS IOVIS

TANTIVM.

MARS MARTIS CIRCVLVS
IVPITER IOVIS CIRCVLVS
SATVRNVS SATVRNI CIRCVLVS

Five Metrics

Excerpted from *Enlightenment Now* | Three

"You're way less likely to die on the job. Every year, 5,000 people die from occupational accidents in the U.S. But in 1929—when our population was less than two-fifths the size it is today—20,000 people died on the job. People back then viewed deadly workplace accidents as part of the cost of doing business. Today, we know better, and we've engineered ways to build things without putting nearly as many lives at risk."

CIRCVLVS IOVIS

TANTIVM.

MARS MARTIS CIRCVLVS
IVPITER IOVIS CIRCVLVS
SATVRNVS SATVRNI CIRCVLVS

Five Metrics

Excerpted from *Enlightenment Now* | Four

"The global average IQ score is rising by about 3 IQ points every decade. Kids' brains are developing more fully thanks to improved nutrition and a cleaner environment. Pinker also credits more analytical thinking in and out of the classroom. Think about how many symbols you interpret every time you check your phone's home screen or look at a subway map. Our world today encourages abstract thought from a young age, and it's making us smarter."

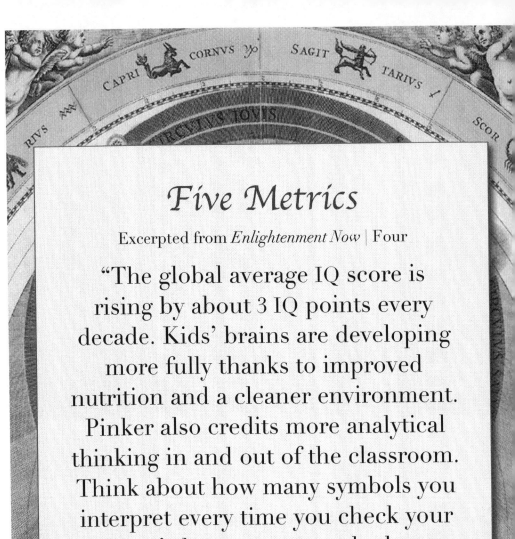

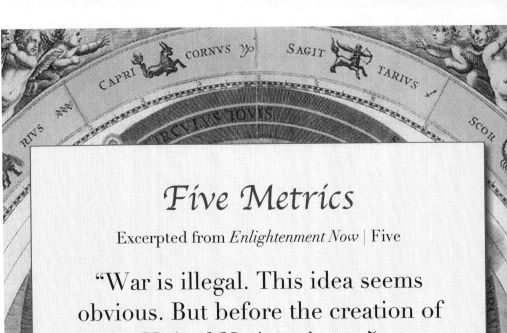

Five Metrics

Excerpted from *Enlightenment Now* | Five

"War is illegal. This idea seems obvious. But before the creation of the United Nations in 1945, no institution had the power to stop countries from going to war with each other. Although there have been some exceptions, the threat of international sanctions and intervention has proven to be an effective deterrent to wars between nations."

17

Of course, there are other rubrics that we sometimes gloss over, such as the doubling of life expectancy at birth from 1700 when it was roughly around 35 years compared to today which is around 70 years.

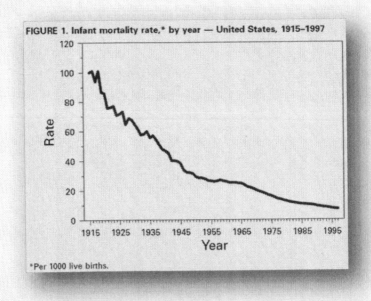

FIGURE 1. Infant mortality rate,* by year — United States, 1915–1997

*Per 1000 live births.

Couple this with the exponential change in the dangers in pregnancy and childbirth for women. Statistically speaking, today 15 mothers die per 100,000 live births, whereas in the 1700s the rate was nearly 10 times that number.

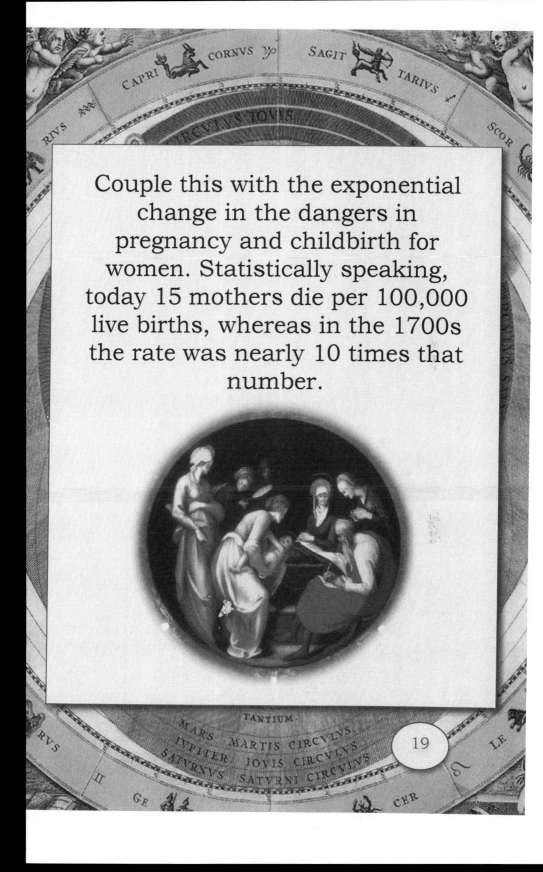

One of the more dramatic improvements over the last 150 years is how many people died from famine in the latter part of the 1800s--with the peak being 142.6 per 100,000 in the 1870s--contrasted with the early 2000s where it is now less than 5 per 100,000.

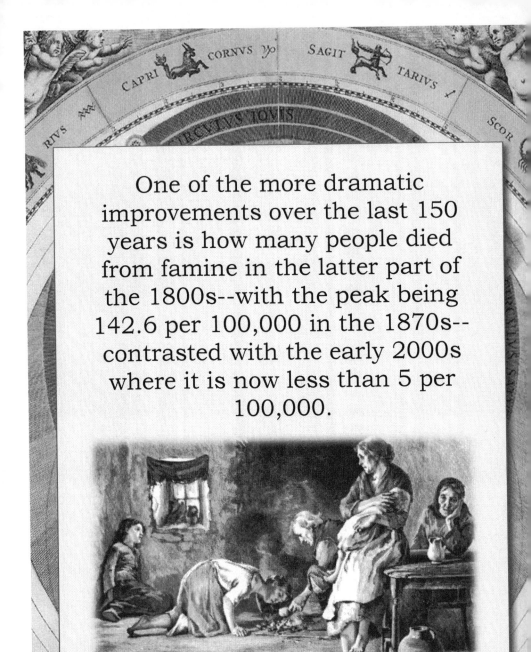

With the progress of modern medicine, many diseases have either been greatly reduced or almost altogether eliminated. Alexander Fleming's discovery of the antibiotic effect of penicillin, for instance, is estimated to have saved the lives of eighty to two hundred million individuals across the globe.

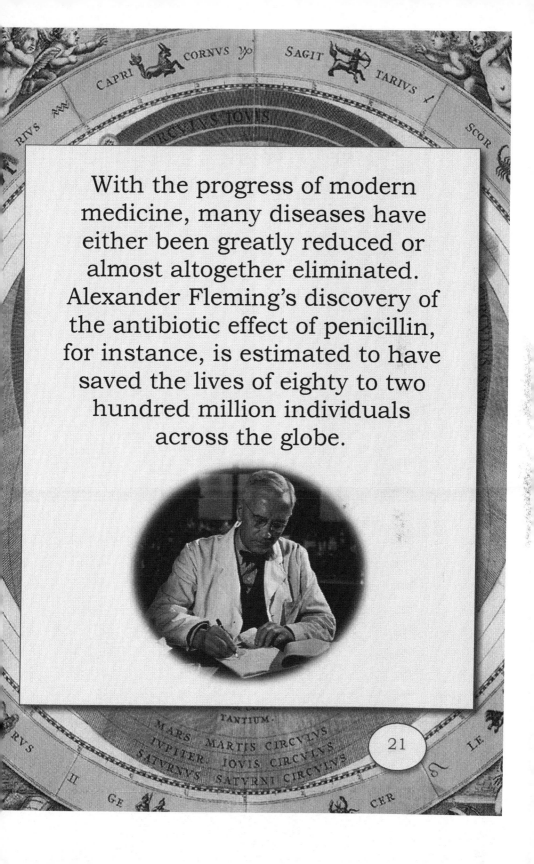

As much as I can readily concur with many of Pinker's salient points, there is no getting around the obvious fact that millions of human beings are still living in horrid conditions and suffering immensely.

"To live is to suffer, to survive is to find some meaning in the suffering."

-- *Nietzsche*

Buddha's great existential observation (and the first Noble Truth in Buddhism) still remains a universal: life is *dukkha*.

It is perhaps no exaggeration to state that to be alive means to suffer, even if we may be lessening some of our ailments and pains over the last few centuries.

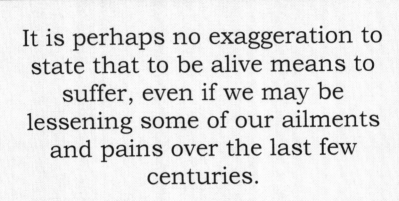

What is the Noble Truth of Suffering? Birth is suffering, aging is suffering, sickness is suffering, dissociation from the loved is suffering, not to get what one wants is suffering: in short the five categories affected by clinging are suffering.

But I think there is a greater message underpinning Pinker's thesis that should not be neglected. Too many of us have the mistaken belief that we lived better lives in the past than we do today. As Bill Gates insightfully pointed out in his recent conversation with Steven Pinker about his research and newest book, "The most stunning thing for me—ever—is this disconnect between the actual improvement and people's view that, maybe the past was better, maybe the future is going to be worse."

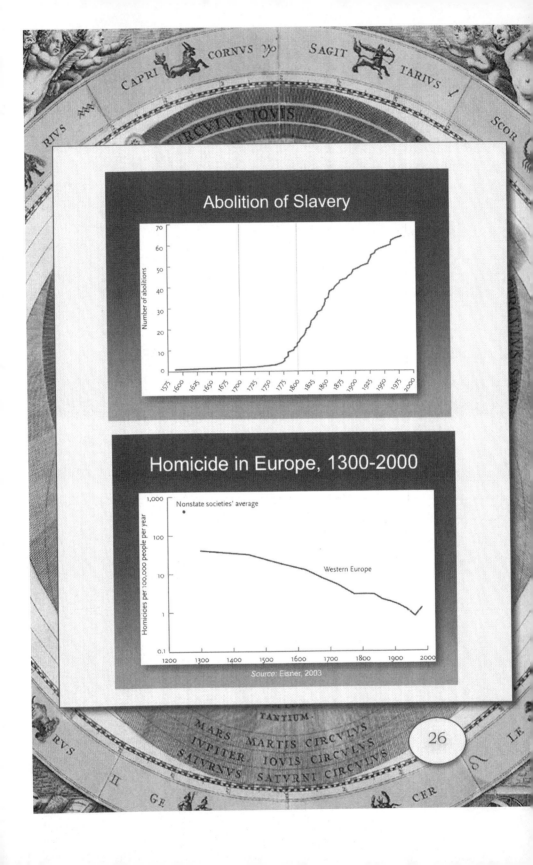

Abolition of Slavery

Homicide in Europe, 1300-2000

Source: Eisner, 2003

26

While it is dangerous to include everyone in the sweeping generalization that life is getting better, particularly because certain individuals, certain families, and certain groups, may have much worse lives now, it is still nevertheless true that for the vast majority life is easier to survive and flourish than in years prior.

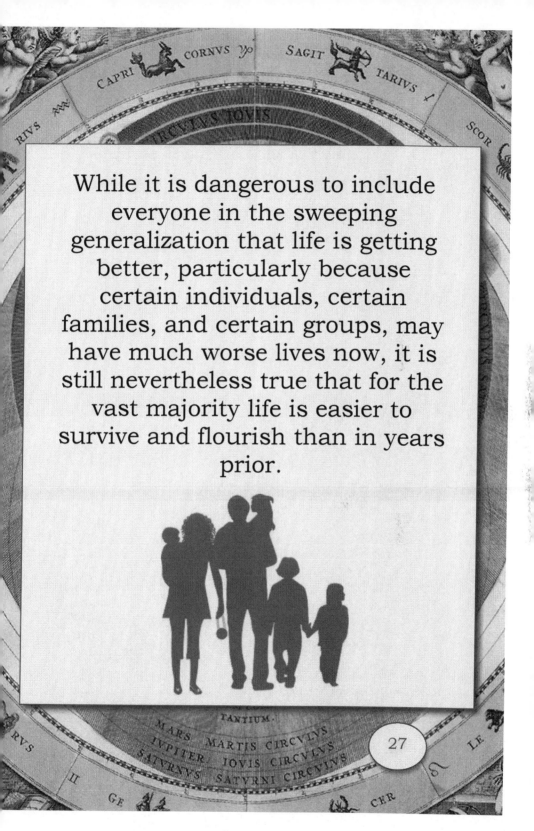

Yet, regardless of whether we agree or disagree with Pinker's rosier outlook, I do think that we can employ comparative histories and sociologies to make us better appreciate our own particular circumstances.

A third way of understanding Comparative Sociology :

- Comparative Sociology is an Ideal Type of Sociology
- Comparative Sociology is a Process
- Comparative Sociology is a Critique

In some ways, Pinker's real agenda is to wake us up from our misplaced nostalgia for all things long past (falsely imagining how it was) and realize how much we have "progressed" even if we use that very term selectively.

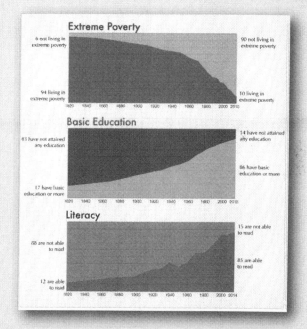

The <u>Meta-Person</u> Project

Pinker's studies reminded me of a program I once worked on in the late 1970s and early 1980s when I was teaching at Catholic high schools (Moreau in Hayward, Chaminade in Woodland Hills, and University High in San Diego). Back then I called it the "Meta-Man Project" (today, that would be changed to "Meta-Person"), where we took a person's entire life from birth to death (using 75 years as our benchmark) and partitioned various aspects of what we could expect to happen.

For instance, how many hours of sleep does a person average in a lifetime? 25 years or roughly 9,125 days.

The Sleep Pattern of Animals

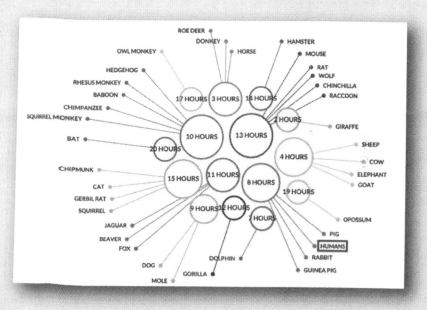

How many breakfasts are we going to consume? 27,000 plus.

How much time is spent urinating? Depending on your bladder, it ranges from 7 to 10 minutes a day or about six months in a life span.

In the book, _Just Odds_, it is estimated that we touch our face from 2,000 to 3,000 times a day (roughly 80 million times in 75 years).

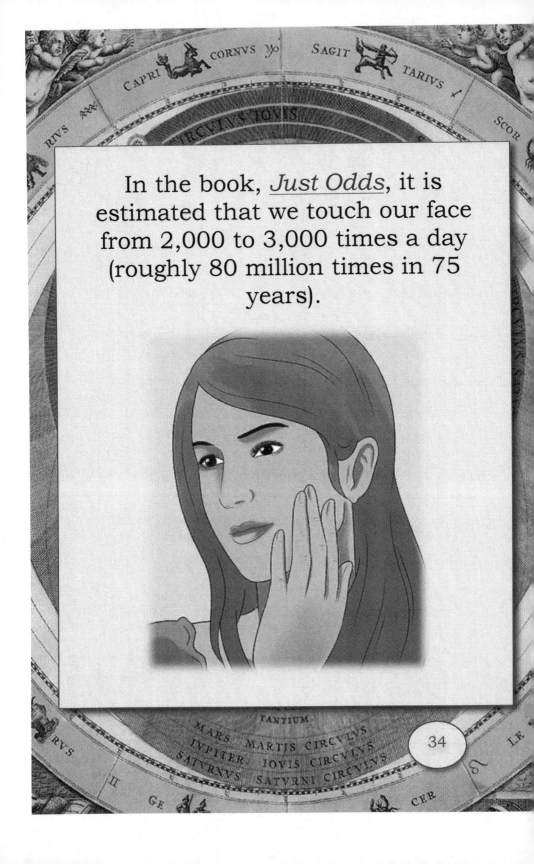

We blink our eyes 15 to 20 times per minute (nearly one half a billion blinks in our life).

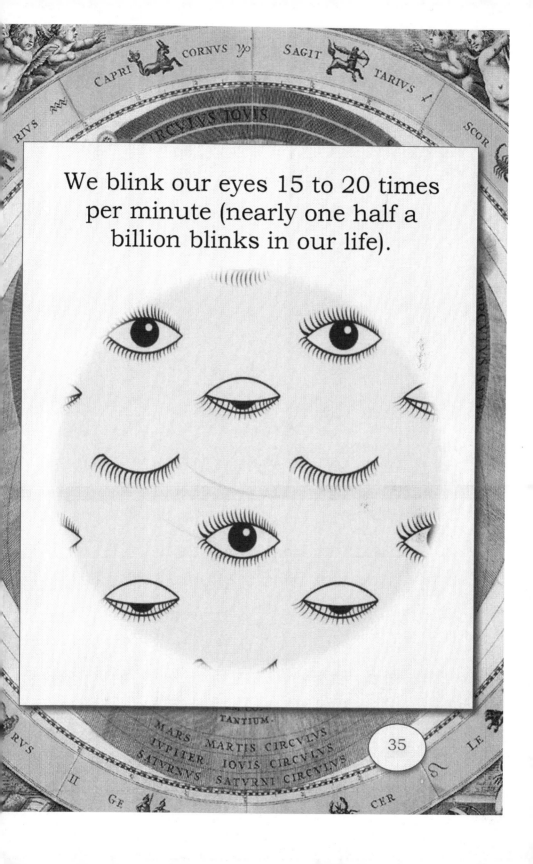

We swallow 600 times a day
(sixteen million swallows till
death do us part).

And according to some estimates
we yawn on average 6 times a day
or 164,250 yawns in a lifetime.

What all these "meta" statistics
do is help us realize what life has
in store for us and establish
upper and lower limit parameters.
Certain things are going to
happen and certain events will
take up your time. Given this
overall picture, it allows one a
clearer conception of what we are
up against and what we wish to
accomplish.

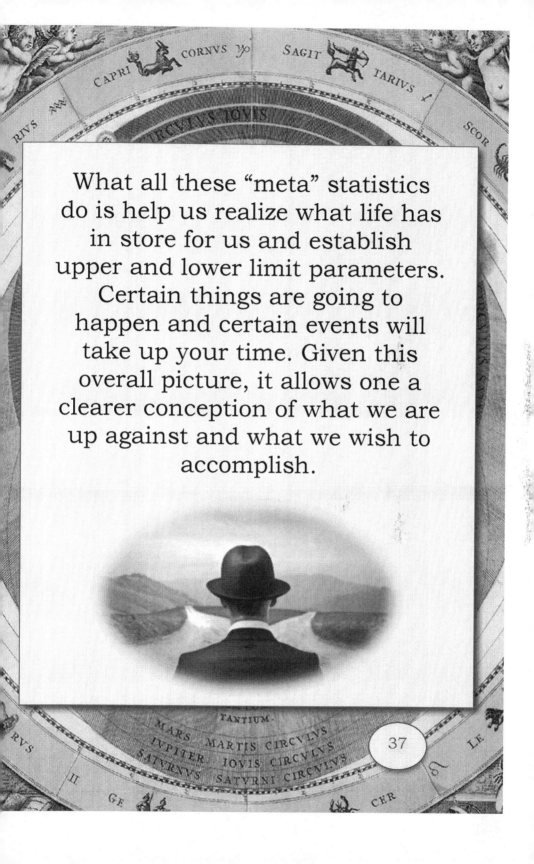

Later when I started teaching courses in science and religion at California State University, Long Beach, we began exploring in depth the theory of large numbers and the dramatic implications of probability theory.

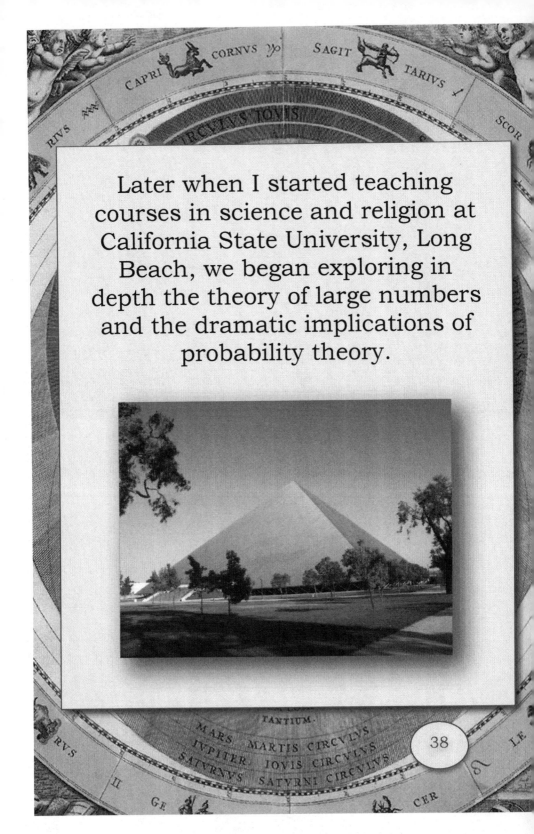

This led to a radical understanding of how much luck is involved in anyone's life and how such can be utilized (in a comparative way) to better appreciate one's position in life.

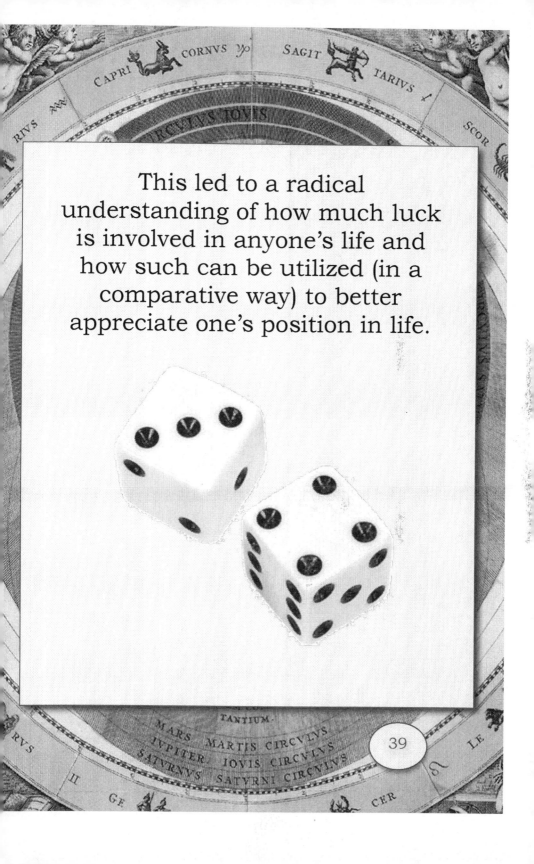

Imagine right now as you yawn away listening to your Professor lecture about the mathematical implications of *Desultory Decussation*, you take out the California Lotto ticket you bought earlier that week at your local 7/11, and realize right then and there that you have just won 10 million dollars.

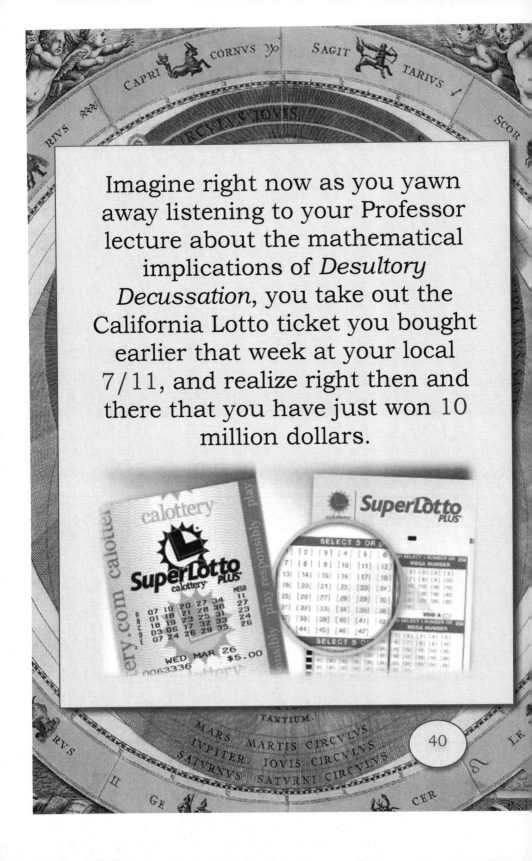

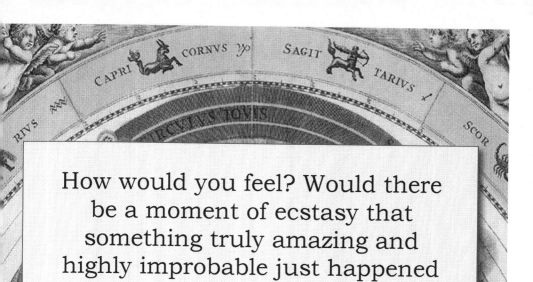

How would you feel? Would there be a moment of ecstasy that something truly amazing and highly improbable just happened to you?

But we have already won so many lotteries just by being alive right now. Yet, we never really do appreciate the wondrousness behind having self-reflective consciousness.

It has been estimated that there have been 108 billion human beings who have lived so far. But what this vast number doesn't convey is just how difficult it is to be one of those rare 108 billion.

All Humans Who Have Ever Lived
(~108 Billion)

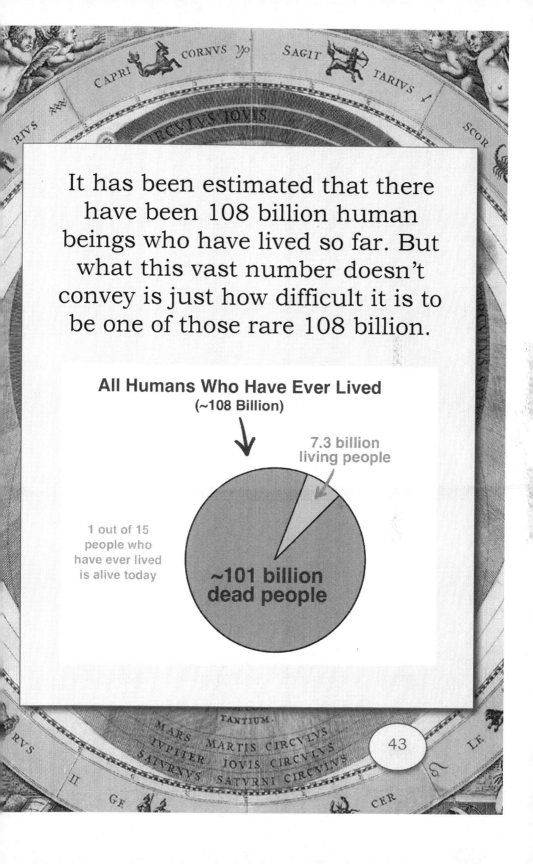

7.3 billion living people

1 out of 15 people who have ever lived is alive today

~101 billion dead people

As Dr. Ali Binazir explains, "The probability of the right sperm meeting the right egg [creating you, in sum] is one in 400 quadrillion"

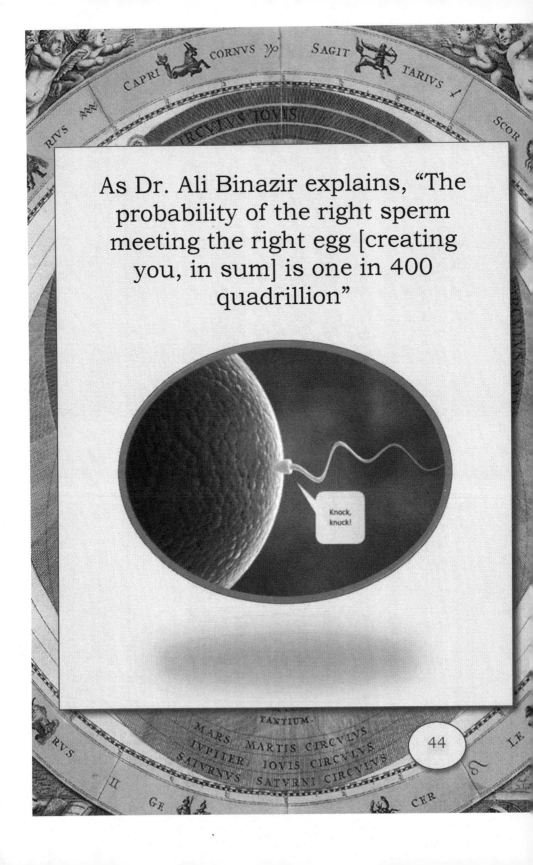

But even that number pales when we stop to realize that each of one our predecessors had to reproduce successfully, since one "screw up" (ironic pun?) means you right now don't exist. Binzair puts that number as <u>1 in 10^{45,000}</u>

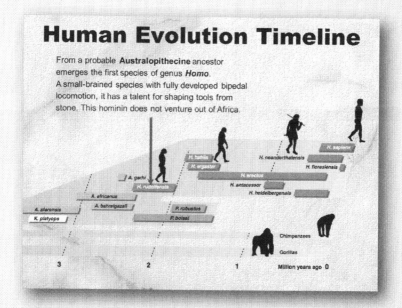

Human Evolution Timeline

From a probable **Australopithecine** ancestor emerges the first species of genus *Homo*. A small-brained species with fully developed bipedal locomotion, it has a talent for shaping tools from stone. This hominin does not venture out of Africa.

So you have already won a "life" lottery that is truly mind boggling, and makes getting ten million dollars akin to winning a penny, since the "probability of being born is one in $10^{2,685,000}$."

> "To distill so specific a form from that chaos of improbability, like turning air to gold... that is the crowning unlikelihood. The thermodynamic miracle . . . For you are life, rarer than a quark and unpredictable beyond the dreams of Heisenberg. . . ."
>
> --Alan Moore, *Watchmen*

Of course, such ad hoc numbering is itself a bit of a ruse since only those who are already alive can play such a parlor game, but nevertheless it does indicate something we should never forget: *we are luck incarnated.*

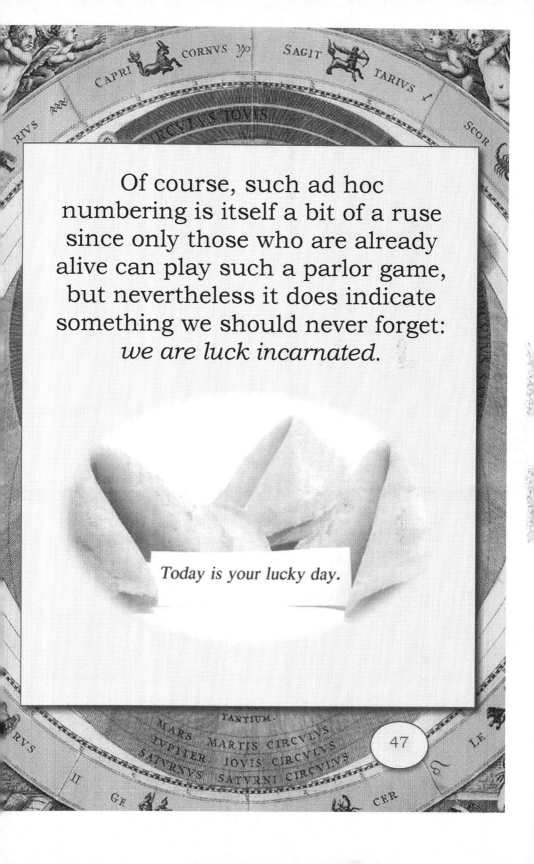

Today is your lucky day.

And thus we should become familiar with such numbers, if only to better understand the lives we do have.

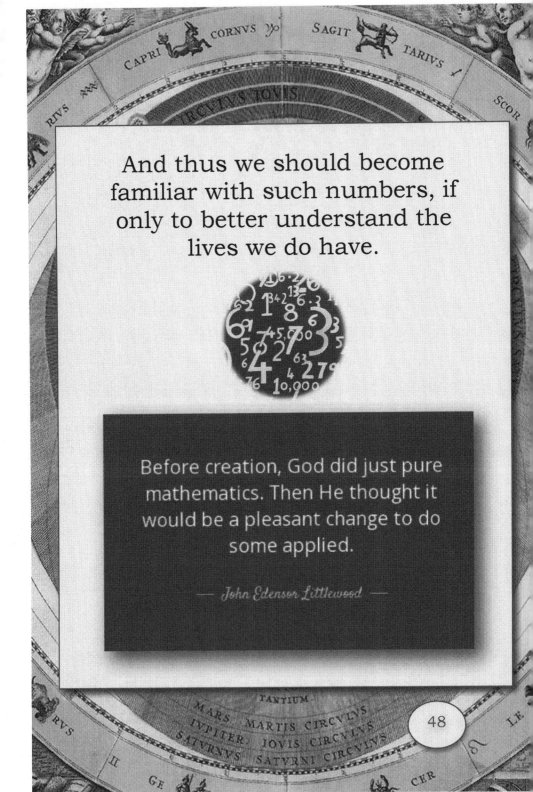

Before creation, God did just pure mathematics. Then He thought it would be a pleasant change to do some applied.

— *John Edenson Littlewood* —

When we were kids in Catholic elementary school we used to complain about having to wear hard wingtip shoes as part of our uniform. We got blisters and they were a pain to play in during recess, especially basketball. However, the nuns reminded us that we were fortunate to even have shoes since many in the world had to walk around barefoot.

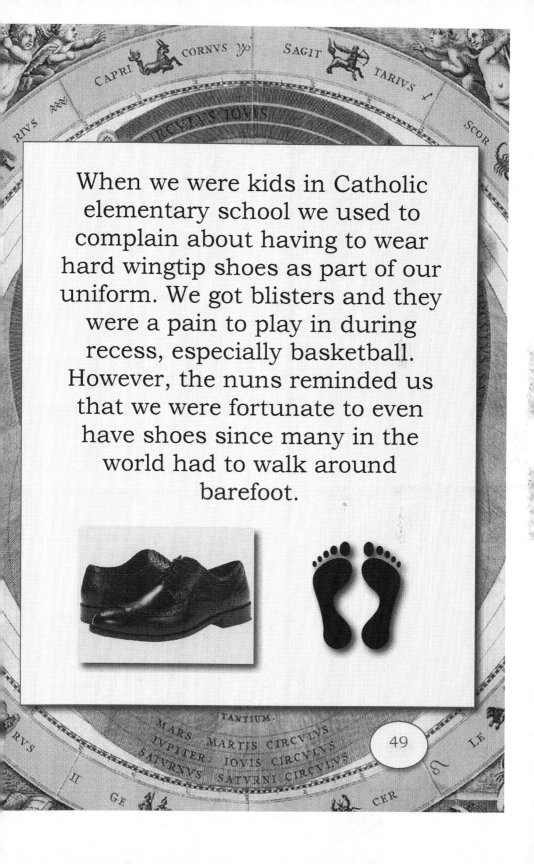

The same kind of lesson was pressed upon us during lunchtime in the cafeteria (which had some truly awful food) where we were forced to eat everything on our plates, since there were millions in India who were starving. Of course, we never did figure out logically how eating all our food actually helped anyone back East.

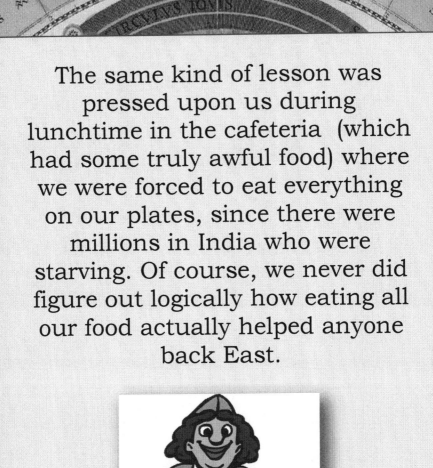

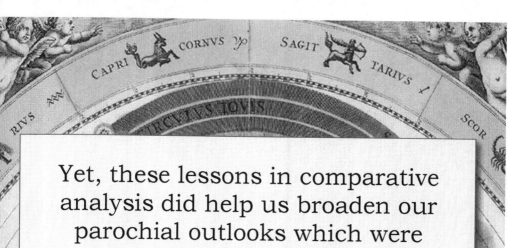

Yet, these lessons in comparative analysis did help us broaden our parochial outlooks which were invariably skewed since we tended to compare our lot in life with other affluent kids in the neighborhood, never quite realizing how truly blessed we had been and how spoiled we really were.

In a way, Pinker is reminding us anew, with a truckload of stats to back up his argument, that we must always see the larger picture in order to properly adjudicate our place in life.

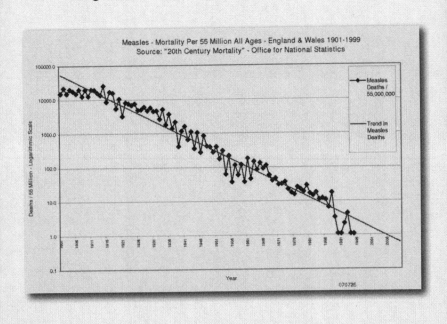

Measles - Mortality Per 55 Million All Ages - England & Wales 1901-1999
Source: "20th Century Mortality" - Office for National Statistics

Once in my Critical Thinking class at Mt. San Antonio College we checked out a website called *globalrichlist.com* to see how the above average minimum wage worker in California compares to others around the world.

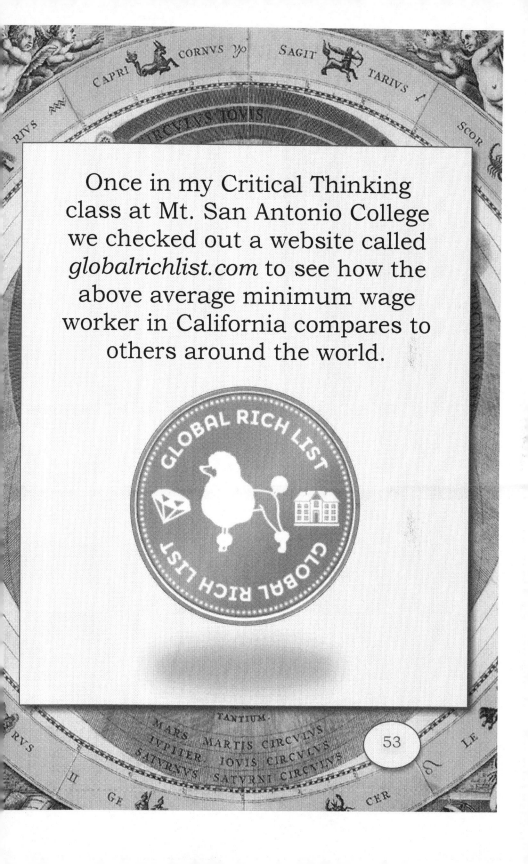

We were all shocked to realize that if you earn 40,000 dollars a year you make more money than 99 percent of all other humans on the planet.

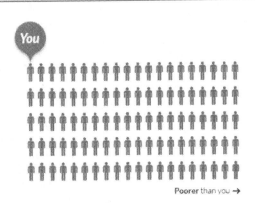

You

Poorer than you →

You're in **the top**

0.57%

richest people in the world by income.

That makes **you** the

33,982,065th

richest person on earth by income.

But generally nobody making 15 dollars an hour in Los Angeles feels like a lottery winner, even though they truly are, given how the rest of the world lives.

The vision of who we are is bounded by the scope of our models with which we compare and contrast ourselves. If we would widen our vistas we would soon realize how truly fortunate we are and perhaps we would see that it all comes down to changing one's perspective in order to make ourselves happier or more content.

For example, putting housing and transportation aside, the average worker today in America has more purchasing power than those living in the 1950s.

When I was a kid we never imagined that in the future there would be various restaurants and fast food joints around the world offering such things as "refillable" soft drinks or self-serve ice cream dispensers where you could go back as many times as you want for the same price.

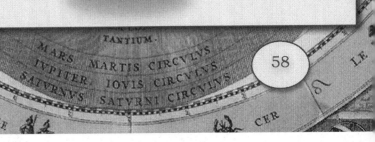

Indeed, so unimaginable was this very idea, that I remember my first grade teacher, Sister Susan, describing heaven as a place where you get as many Cokes and ice cream as you want. Little did we realize then at 7 years old that within two decades Sister Susan't "heaven" would descend on earth in the name of *McDonald's*.

It all comes down to desire and the context in which our wishes arise and play out. Naturally, this doesn't obviate the horrors that plague our life on planet earth, but it does show us a way to temporarily find ways to feel a bit better about life and the odds against it.

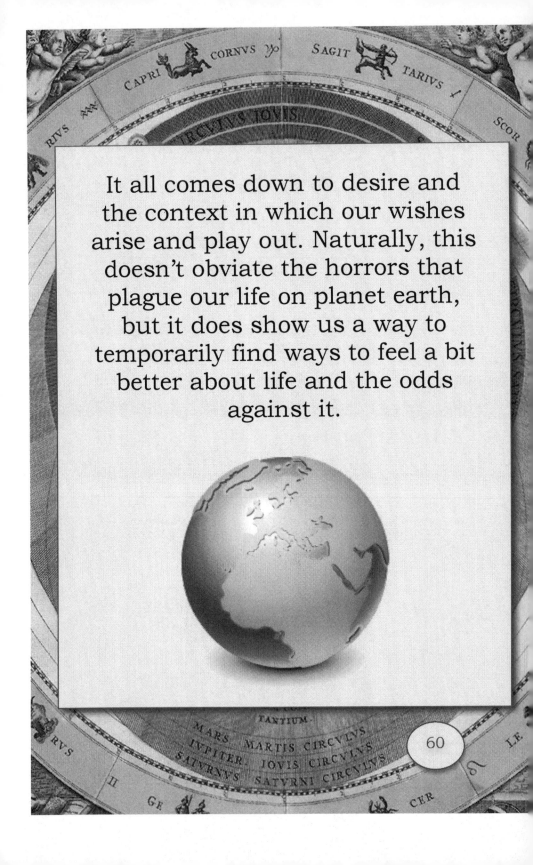

My late friend, Peter McWilliams, famous for writing a number of best selling self-help books, once titled one of his longer tomes, *You Cannot Afford the Luxury of a Negative Thought*, arguing persuasively that since our time here is in short supply we shouldn't waste it away by occupying our minds with ideas that cause us harm.

Although it may be inaccurate to say that Steven Pinker is science's version of Norman Vincent Peale, the famous American minister who championed "positive thinking," his book _Enlightenment Now_ does indeed encourage us to see the present and the future in a much brighter light and not to get bogged down with false memories of some past golden age which never really did exist. According to Pinker's thesis, the best time is precisely now. The only thing holding us back is waking up to it.

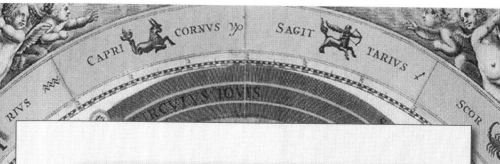

"To be successful is to be helpful, caring and constructive, to make everything and everyone you touch a little bit better."

~Norman Vincent Peale

Lest we forget, we are the lucky 108 billion who been favored to be born human. We are, in sum, winners in the Darwinian lottery of life.

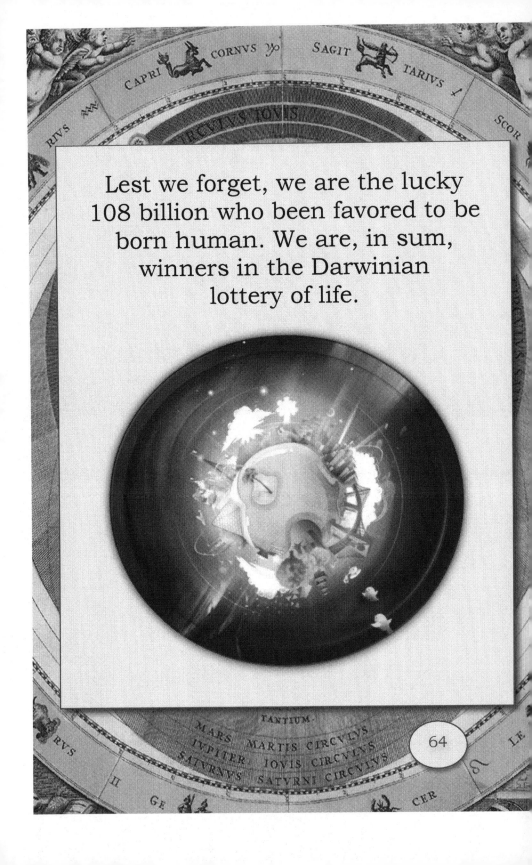

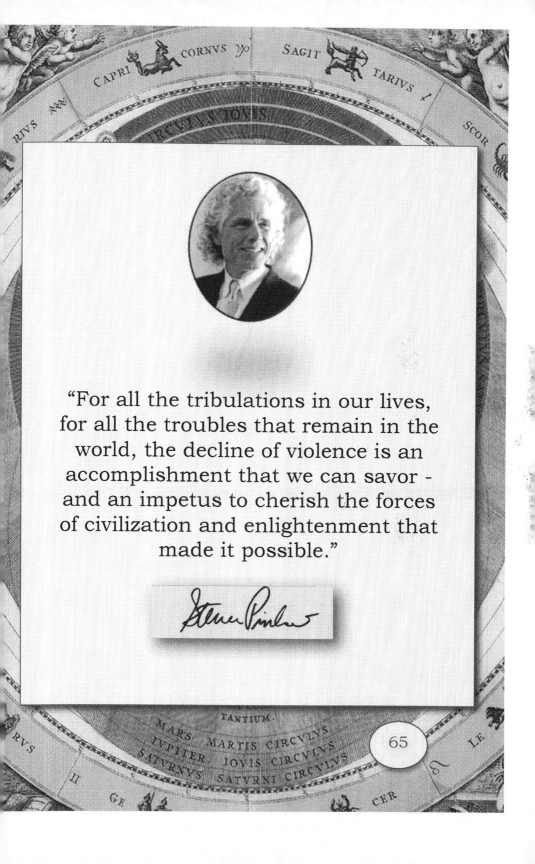

"For all the tribulations in our lives, for all the troubles that remain in the world, the decline of violence is an accomplishment that we can savor - and an impetus to cherish the forces of civilization and enlightenment that made it possible."

"The world is getting better, even if it doesn't always feel that way."

--Bill Gates

Made in the USA
Lexington, KY
27 March 2018